Feeding Honeybees Throughout the Year

David MacFawn

Feeding Honeybees Throughout the Year
Copyright © David MacFawn

Published 2025 by
Northern Bee Books,
Scout Bottom Farm,
Mytholmroyd,
West Yorkshire
HX7 5JS (UK)
Tel: 01422 882751
Fax: 01422 886157
www.northernbeebooks.co.uk

ISBN 978-1-914934-94-0

Design and artwork DM Design and Print

Foreword

Beekeepers spend considerable time ensuring their colonies have adequate nutrition. In fact, beekeepers often consider nutritional stress one of the most significant threats to honey bee colony health. Thus, it is imperative that every beekeeper be able to recognize nutritional stress associated with nectar and pollen deficiencies and know how to remedy these issues as they arise. Fortunately, Dave MacFawn developed a guide for these very purposes. In this book, Dave shares multiple strategies, tips, and tricks that beekeepers can use to ensure their colonies are well-nourished. Even the most experienced beekeepers will find something they can use to manage their colonies' nutritional health.

Jamie Ellis, PhD
Gahan Endowed Professor
Honey Bee Research and Extension Laboratory
Entomology and Nematology Department
University of Florida

Larry Coble (EAS Master Beekeeper) and Dr. Tim Liptak (EAS Master Beekeeper) contributed to the overall content definition. Special thanks to Ann Chilcott for reviewing the manuscript and making helpful comments. www.beelistener.co.uk

Contents

Figures

Introduction to Feeding

The overall colony management goals might be missed unless a comprehensive approach to feeding is taken. Making a feeding plan for the entire beekeeping year ensures a proactive strategy that will be better for the colony than a reactive one made after problems have arisen. Feeding your bees for survival is critical if they are out of food. Sugar syrup feeders and pollen feeders are important for feeding. Each type of feeder has its place. In the Southeast United States, the pollen feeder has replaced pollen patties due to small hive beetles (SHB) *Aethina tumida* reproducing in pollen patties.

Each type of feeder has its use, benefits, and disadvantages. The pail feeder is the most versatile for warm and cold weather. It is also an inexpensive way to feed syrup. Yard feeders are even less expensive but have some disadvantages (covered later). Pollen feeders are also important for continued brood production during pollen dearth especially in autumn with the understanding the bees may not last all winter. Feeding dry pollen substitutes in pollen feeders alleviates some small hive beetle issues inside the hive.

The author is from the middle part (Mid-State) of South Carolina, United States which is in the southeast part of the US.

Chapter 1:
Sugar Syrup Feeder Equipment

Figure 1: A Two-Gallon Pail Feeder on Top of the Equipment Stack Larry Coble Photo

Only pure white crystalized sugar should be used. Confectioners' sugar with starch should not be used. It causes dysentery in the bees. Also, any dark brown or light brown sugar or raw sugar should not be used.

Pail feeders are inexpensive (about $10 for a two-gallon). However, they do require an inner cover. The pail feeder is inverted and the pail sides are squeezed and released to create a vacuum so the feeder will not leak. Then the feeder is placed over the Porter bee escape hole on the inner cover. Pail feeders are easy to fill, transport, and clean. Most beekeepers place a deep super around the pail feeder to prevent hive issues with the weather. Tiny holes punched in the lid or a screen allow the bees to access the syrup. If you use a migratory cover without an inner cover as your standard configuration, you will need to use an additional inner cover to place the pail feeder on.

Figure 2: A Frame Feeder on the Left Photo Larry Coble

Frame feeders, also known as division board feeders, work well in warm weather. However, in cold weather, the bees may not be able to access the sugar syrup. Frame feeders are easy to fill when the feeder is placed on the side of the brood chamber or super. The super above is merely slid over, the frame feeder filled, and then the super slid back into place. A float, such as a twig, popsicle stick, or other material needs to be placed in the frame feeder to keep bees from drowning. Cleaning the frame feeder may be tricky since it needs to be removed from the hive.

Figure 3: Danny Cannon's Homemade Jar NUC Feeder on left and 10 Frame 4 Jar Feeder Top on Right David MacFawn Photo

Jar feeders work well in warm or cool weather in the southeast. In much of the Southeast United States, it rarely gets below about 25 degrees Fahrenheit (-3.8 degrees C). Sugar syrup will typically not freeze in much of the Southeast. Bees can access the syrup in warm or cool weather. When the bees move upward through the equipment stack in the winter, and they exhaust their honey stores and reach the feeder, the bees can huddle under the feeder and access the sugar syrup. Jar feeders are inexpensive, the jars can be transported and cleaned easily. It should be noted that glass jars may break in the bee yard causing a safety issue. Thick plastic jars are recommended; thin plastic jars will collapse.

3

Figure 4: A 10-Frame Hive Top Feeder Dadant Photo

A ten-frame hive top feeder (also called miller feeders) with floats costs around $35.00. A two-gallon pail feeder costs around $10.00. An inner cover costing around $15 is also needed, but the beekeeper only has to spend $25 on this set-up. Therefore, a hive top feeder configuration is about $10.00-$15.00 more expensive than a two-gallon pail feed configuration. If you only have a few colonies, it may not matter. A hive top feeder is easier to refill than a pail feeder which results in less management time being spent on the task. With hive top feeders, the bees access the syrup via the opening between the reservoirs. Mold in both will need periodic cleaning. Hive top feeders will hold greater than 2 gallons (7.57 liters), and a pail can hold about 2 gallons. The feeder size is important if the beekeeper has outyards. Larger feeders are preferred for outyards so you do not have to make as many trips.

Yard or open, feeders should be at least approximately 200 feet (61m) away from the apiary to reduce the risks of robbing. A yard feeder may be as simple as a bucket with straw so the bees do not drown in the sugar syrup. However, the issues with yard feeders are:

- Weak colonies may not get their fair share of the sugar syrup
- They may spread diseases
- The bees cannot access the sugar syrup in cold weather when the bees do not fly (typically less than 48F (8.9C) to 50F (10C))

However, yard feeders have their place regarding time efficiency and reduced labor.

Figure 5:Larry Coble's Five Gallon Yard Feeders David MacFawn photo

Figure 6: Holes in Top of Yard Feeders. Larry Coble Photo

Use a small sharp nail for the holes rather than a 3/32" (2.3mm) drill bit. The bees do not seem to propolize the holes shut.

Figure 7: Boardman Entrance Feeder. David MacFawn photo

Boardman entrance feeders should only be used to dispense water. Feeding sugar syrup via a boardman can cause robbing. Also, feeding sugar syrup via a boardman in the winter results in the bees not being able to access the syrup if the bees are clustered. Visible clustering occurs at 57 degrees F (13.9 degrees C). Boardman feeders are inexpensive and cost only around $8 plus the jar cost.

Chapter 2:
Pollen and Pollen Substitute Feeders

The importance of a varied pollen source is very important for the bees' health and disease prevention. Bees require pollen/protein for growth, development, and immune function. A lack of multiple pollen sources can weaken the bees' immune systems, making them more vulnerable to diseases.

When looking at the colony's stored pollen it should see various colors, not one color. Monoculture crops, habitat loss, or adverse weather conditions, may reduce a colony's varied pollen sources. A varied pollen source will add multiple nutritional sources to the gut biome and its importance physiologically. We are only now perfecting pollen substitutes and still have a way to go in researching the constituent ingredients.

Pollen Feeders

In some locations, there may be a seasonal pollen dearth. A pollen dearth is especially critical in autumn when the winter bees are developing. Lack of pollen will reduce brood production even if you have plenty of nectar/honey. It takes pollen and honey to produce young bees. Hence, it may be viable and important to feed pollen substitutes to stimulate brood production. It should be noted, that University of Florida studies indicate the bees are seeking the sugar in pollen substitute and may not store the pollen. Other studies indicate that feeding pollen substitutes in autumn, develops summer bees rather than winter bees. This means pollen-fed bees may not last as long as bees developed without a lot of pollen.

Several methods are available. The beekeeper can feed pollen patties or dry pollen in feeders. Pollen patties have lost favor in the Southeast due to small hive beetles (SHB), reproducing in the pollen patties. Dry pollen in pollen feeders has replaced pollen patties as the method of choice in the Southeast. A pollen feeder may be as simple as a pail turned 90 degrees on its side and nailed on a tree (Figure 8) or a more elaborate pollen feeder (Figures 9 &10).

Figure 8: Two-Gallon Pail Pollen Feeder David MacFawn Photo

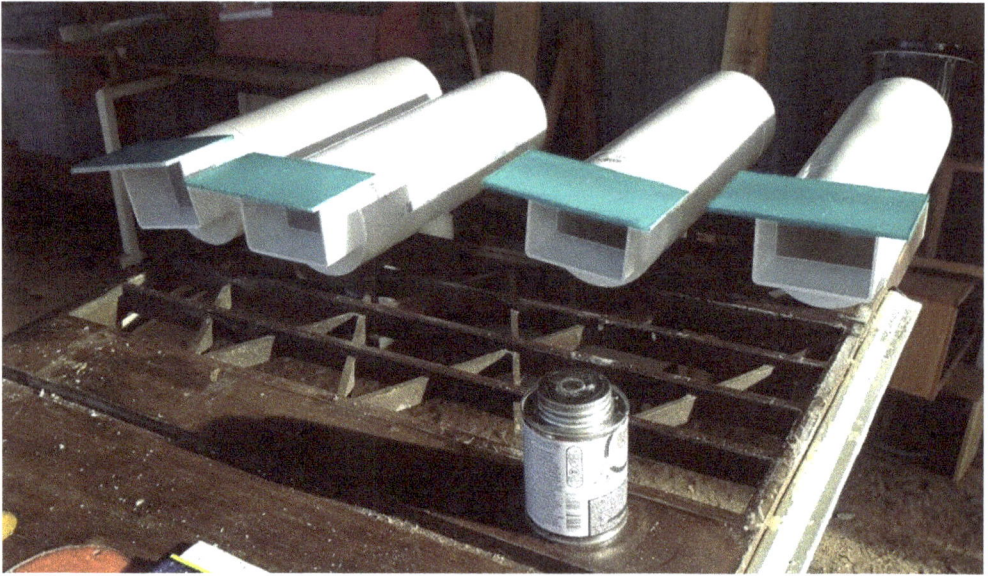

Figure 9: Homemade Pollen Feeder Larry Coble Photo

Figure 10: Homemade Pollen Feeders in the Field Larry Coble Photo.

Another way to feed dry pollen substitutes is to place a small amount on top of an inner cover. However, you need to carefully watch out for dampness. If the pollen substitute gets wet from hive moisture it may attract SHB. Putting pollen substitutes on top of an inner cover will help colony pollen shortage in cold weather when the bees cannot fly.

Pollen patties should be placed immediately above the brood nest. If you use only a small amount, that can be consumed by the colony in a couple of days, there is less risk of attracting SHB. However, each hive has to be opened which requires a lot of time.

It is currently considered better to feed trapped pollen from your hives, than pollen substitute. Your pollen typically has more nutrients than pollen substitutes. Purchased pollen may contain diseases.

Chapter 3:

Water

In the Southeast United States, feeding water may make sense when it gets into the upper 90 degrees F (32.2 degrees C) to 100 degrees F+ (37.8 degrees C) temperatures. A Boardman entrance feeder can be used. Also, feeding 10%-20% sugar syrup in a pail feeder helps maintain hive weight and hydrate the colony. The bees sprinkle water droplets throughout the hive and fan at the entrance to cool the hive.

Chapter 4:
Seasonal Approaches

After the spring nectar flow in South Carolina, wait two to three weeks before feeding bees during the summer dearth. This will allow the colony to reduce egg-laying and produce more bees. The spring nectar flow in Mid-state South Carolina is typically over after the first week or two of June. This means, that if the colony is light in weight, it should not be fed until around the first of July. If you are looking to reduce your sugar expense, you are better off leaving at least a medium super of honey on the hive.

Feed in the autumn rather than in the spring to make sure the colony has enough honey till the spring flow starts end of March. However, in some years you may need to feed in the spring depending on the year. Feeding in the spring results in bee buildup and swarming. However, if you want to split the colony in the spring, feeding may be in order.

Feeding simulates a nectar flow resulting in the colony raising more bees or maintaining a large number of bees. If the beekeeper does not want to maintain the expense of a large number of bees during the summer dearth, leaving honey on the hive to reduce cash outlay may make sense.

In much of the Southern United States, including most of South Carolina, we can feed sugar syrup from inverted pail feeders all year round. It rarely gets into the 20 Degree F range. Candy boards can be used but are more of a cold climate feed. (www.mannlakeltd.com > blog > how-to-make-candy-board)

During warm weather, the feeders should be cleaned every week, to a week and a half. During cold/cooler weather the feeders can be cleaned every 2-3 weeks. It should be noted when the weather turns cooler/cold, in November in South Carolina, the bees will leave the syrup in the feeders and not store it in the hive. In South Carolina, beginning around mid-September, the bees store the syrup in the hive around the brood nest and feed super.

Thin 1:1 syrup, by volume or weight, should be used in the spring for bee build-up. Thicker 1 syrup is typically used in autumn. Colonies typically stop taking sugar syrup when a nectar flow occurs. Feeders should be removed during nectar flows if you plan to harvest the honey to ensure no contamination/adulteration of the honey.

After I harvest honey my last honey, which is typically at the beginning of June, I configure the hive in their winter configuration, which is a deep and medium food chamber. This allows the bees plenty of time to organize the colony. It should be noted in South Carolina, the main nectar flow in most areas is the beginning of April to the second week in June in the Mid-State South Carolina area.

Chapter 5:
Feeding is Local

Feeding both sugar syrup and pollen is local. Vegetation varies within a region and across regions. This means you need to assess a bee yard's carbohydrate and pollen needs. This may take a year or two.

When to feed for spring buildup and possible splitting depends on your latitude. In the Southeast United States, you may need to feed in January for a March split and April nectar flow. In New England, the nectar flow is much later so you would feed later in the spring unless a colony is light on food stores.

Chapter 6:
Issues with Moisture and Feeding

The hive configuration is important for winter survival. Thick insulated top covers will help ensure winter moisture from honey consumed will not condense on the inside top cover, with the resulting cold water raining down on the cluster.The ideal situation is when moisture condenses on the outside walls. Bee can collect water from here to thin their honey before consumption. If an inner cover is used, it should not contain a hole in the rim.

Feeding sugar syrup in the winter when temperatures are below freezing may cause issues with moisture. Candy boards can be used at these times.

Figure 11 Condensing Insulated Hive Cover David MacFawn Photo

Chapter 7:
Goals & Methods for First-Year Colonies

The best time of year to establish a colony is at the beginning of the spring or summer flow (in the colder latitudes), so the colony will have time to get established. If purchasing a package of bees or a five-frame NUC, the first-year feeding goal is to get the colony firmly established. Upon installation of a package, it will need feeding to help it get established. If establishing the package on frames with foundation, the goal is to get the bees to draw out the comb. It is usually difficult to get the bees to draw comb with sugar syrup; it usually takes a nectar flow. It takes nectar for the bees to consume to produce fresh wax from four pairs of wax glands on the underside of their abdomens. If drawn comb is available, it will save a lot of time. The bees will be able to refurbish and clean the comb, and the queen will be able to start laying eggs quicker.

The initial colony should be fed until they start collecting nectar, and you do not have any inclement weather where they cannot fly to collect nectar, pollen, and water. The bees should be fed 1:1 (one part sugar to one part water) sugar syrup. In the Southeast United States, there is a summer dearth which means that you will most likely need to feed during summer/these months. One of your goals is to have enough drawn comb for the colonies to store winter "honey."

Bees use the beeswax on the coated plastic foundation to draw out the comb. If you get single-coated wax plastic frames, you should consider rubbing the plastic foundation with a block of beeswax to put a heavier coat of beeswax on the foundation. Wired beeswax foundation does not have issues with the bees drawing out the comb. However, wired beeswax foundation frames will "blow out" , i.e. be destroyed, in the radial and tangential extractor quickly.

Chapter 8:
Goals & Methods for Established and Production Hives

Established colonies usually have drawn comb in both the brood chamber and a feed chamber. The brood chamber is usually a deep (9 5/8") (963cm)and the feed chamber is usually a medium (6 5/8") (663cm) or a deep depending on how much honey is required for your local winter. It should be noted, if interested in honey production, you want as small/ shallow a feed chamber as possible so you can remove full supers placed on top of the feed chamber, i.e. the bees do not store more honey than needed for the winter in the feed chamber.

Chapter 9:
Introduction to the Feeding Year

The feeding year runs alongside the bee year. It starts in December when the bees start building up, through the spring (SE US) to summer nectar flow (Northern US). In the autumn, feeding is to ensure the colonies have enough carbohydrates (honey) to get them through the winter. During the winter, the colonies need to be checked to ensure they have enough food to get them through this period. The amount of honey can be assessed by lifting the rear of the hive. Does it feel heavy and of the required amount of weight for the season? In warmer climates, a medium super may be enough honey. In colder climates, you may need a deep super of honey. When the bees cluster around 57 degrees F (13.9 degrees C) they need enough honey for them to consume to produce heat. The cluster will typically move up through the equipment stack consuming honey. If the honey is not within reach of the cluster they may starve. This is why fall feeding, starting in mid-September, is so important. The bees need to locate the honey where the cluster will be in winter as they move up the equipment stack.

Chapter 10:
Winter Feeding

The colony starts building up after the winter solstice in December. Initially, they take little honey and pollen. However, by the end of February in South Carolina the demand has greatly increased until the nectar flow around the first of April. There is usually enough pollen stored in the comb until the bees can gather fresh pollen at the end of January/early February. It takes honey and pollen to produce bees. If you feed syrup the third to fourth week of January, you can split your colonies by the first week or two in March. If queens are not available, a walk-away-split may be in order.

It should be noted that sometimes an inferior queen may result from a walk-away split. A walk-away split queen can be evaluated and replaced if this occurs. Dr. David Tarpy's lab at North Carolina State University indicated, from their queen research, if the capped queen cells are culled exactly five days after splitting, it results in a reasonable queen. Often the bees will choose an older larva to produce a walk-away queen. A walk-away split may be required in South Carolina at the end of February due to a lack of mated queen availability. The split-half with the original queen should continue to build up properly and often you will get a reasonable honey crop. Walk-away splitting results in the half with the new queen obtaining the genetic material from the local area.

When a colony is fed a two-gallon feeder at the end of January, the syrup will last two to three weeks. It may only last one week in March or during warmer weather and heavy brood production.

Chapter 11:
Spring Feeding

The nectar flow in much of the Southeast United States is in the spring. In Midstate South Carolina it is from April to around the first week in June. Feeders should be removed from established honey-producing hives to avoid any possible honey adulteration.

In more northern climates, feeding can continue up to the nectar flow. Once you start feeding, you should continue until the nectar flow to prevent starvation.

Chapter 12:

Summer Feeding Post Harvest

In most of the Southeast United States, there is a dearth after the spring flow. The beekeeper needs to decide if they will leave enough honey on the hive or feed sugar syrup. Most beekeepers believe honey is healthy and reduces cash outlay for sugar. However, honey, especially local honey, is worth more than sugar.

In cooler latitudes, like New England, the nectar flow is in the summer. No feeding is required during the summer nectar flow.

Chapter 13:
Fall Feeding

In the Southeast United States, fall feeding begins in September and continues until around the end of November. In the Midstate area of South Carolina, the bees will start storing sugar syrup and any nectar from a usually light fall flow. The syrup/nectar is stored strategically around the brood nest and feed chamber, organized for consumption during the winter. Queen excluders should be removed so the queen can move up the equipment stack with the cluster as the winter progresses.

When it gets colder (into the 40 degrees F (4.4 degrees C) and 30 degrees F (-1.1 degrees C)), the bees will leave sugar syrup in the feeder and stop storing it in the hive.

Appendix1:
Money and Time-Saving Feeding Tricks and Tips.

For mixing sugar syrup in a small number of feeders, use a paint stirrer that fits into a battery-powered drill. Use warm water to get sugar to dissolve quicker and more completely.

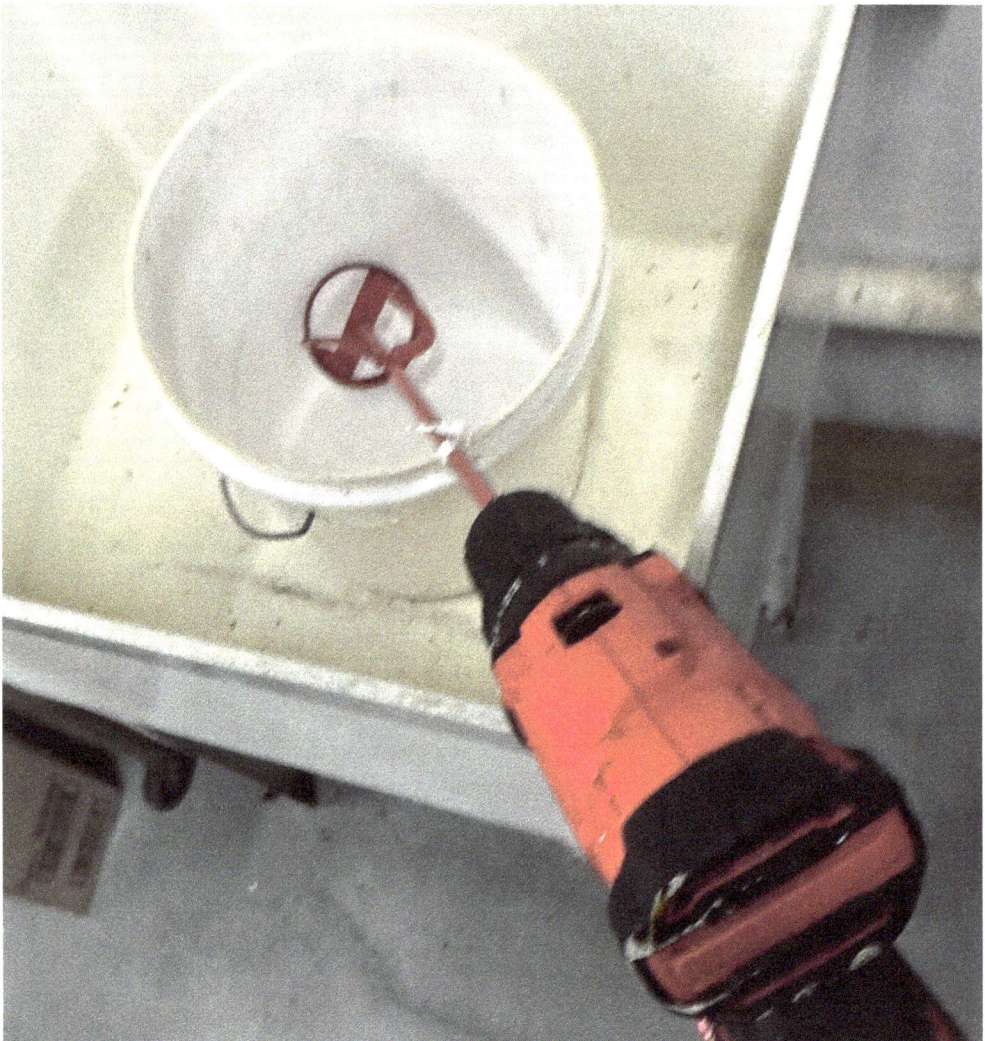

Figure 12 Paint Stirrer in Drill Chuck David MacFawn Photo

Figure 13 Using the Paint Stirrer David MacFawn Photo

Brian Motley: Bulk feeder for a sideline operation.

I used a 30-gallon drum but you can use whatever size you need. I made it so that everything is contained inside so I can put the lid on and bees can't get into the syrup mixture. I used a 1hp sump pump and 3/4" PVC. The ball valve inside is to adjust the flow and the one on the outside controls the fill hose. You could put this in your truck and fill hives in the field. When mixing, you put water in first then the sugar. If you need 2:1, mix 1:1 first then add more sugar. The return line has a 45-degree elbow on the bottom to "stir" the syrup. Warm water and it will mix sugar up quickly. The pump will heat the water as well. It generates heat while running.

Figure 14 Thirty Gallon Sugar Syrup Mixer Brian Motley Photo

Figure 15   Thirty Gallon syrup mixer closed Brian Motley Photo

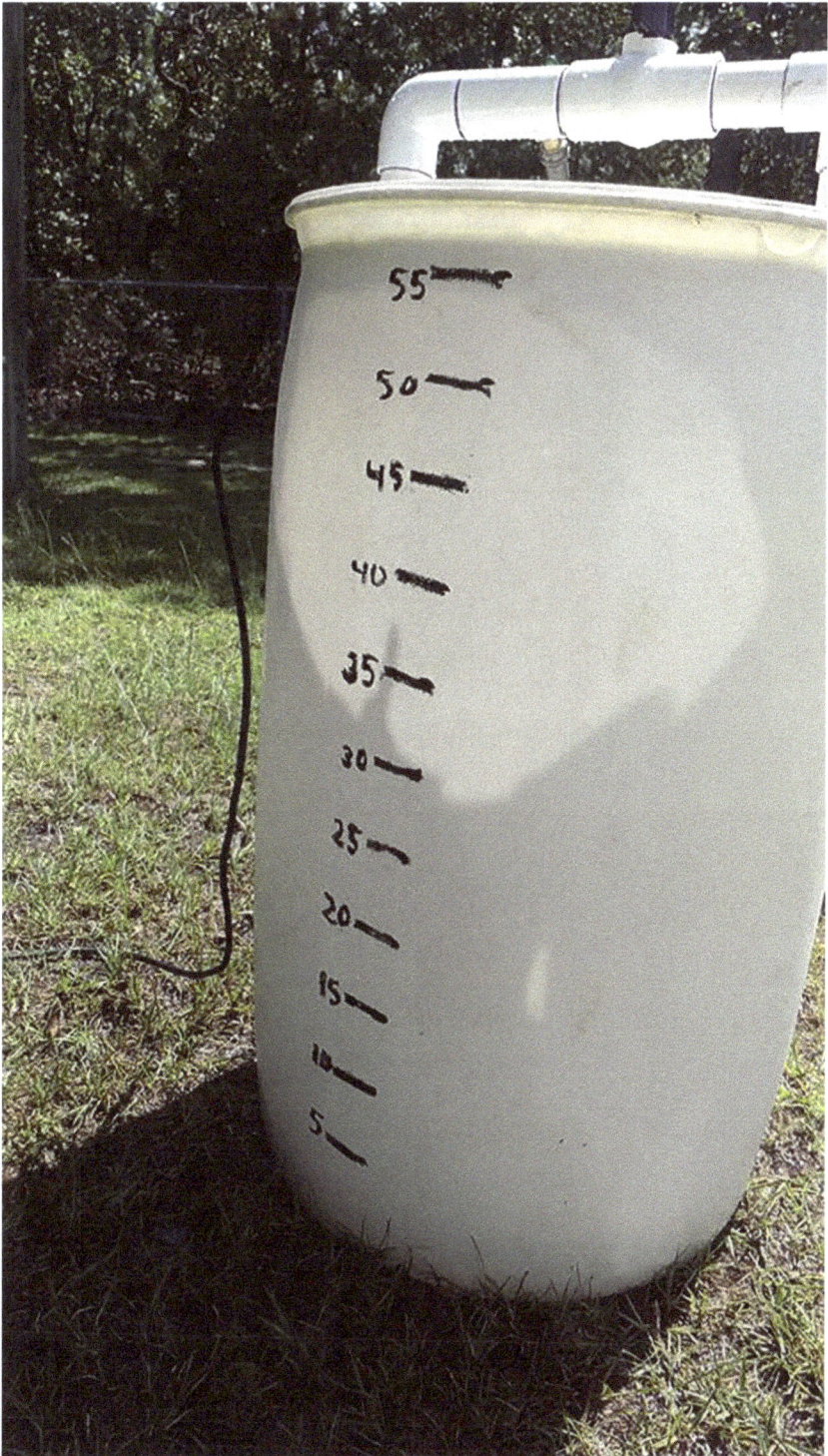

Figure 16 Everything Inside Mixer Brian Motley Photo

Figure 17   Hose Attached to Syrup Mixer, Brian Motley Photo

Figure 18   Inside Mixer Brian Motley Photo

References

1 Mangum, Wyatt A. "Top-Bar Hive Beekeeping: Wisdom & Pleasure Combined," ISBN 978-0-9851284-0-1.

2 Brodschneider, R., Crailsheim, K. Nutrition and health in honey bees. *Apidologie* **41**, 278-294 (2010). https://doi.org/10.1051/apido/2010012

3 **Successful application of anthranilic diamides in preventing small hive beetle (Coleoptera: Nitidulidae) infestation in honey bee (Hymenoptera: Apidae) colonies**

Ethan J Hackmeyer, Tyler J Washburn, Keith S Delaplane, Lewis J Bartlett

Journal of Insect Science, Volume 23, Issue 6, November 2023, 12, https://doi.org/10.1093/jisesa/iead096 06 December 2023

DAVID MACFAWN is a North Carolina Master Craftsman Beekeeper (1997), and an Eastern Apicultural Society Master Beekeeper (2019). He helped found the South Carolina Master Beekeeper Program, and has twice received (1996 and 2020) the South Carolina Beekeeper of the Year Award. He has published over 60 articles in the magazines American Bee Journal, and Bee Culture. David has also written five books, most notably *Applied Beekeeping in the United States*, and has produced twenty You Tube videos. He was a consultant to Bee Downtown, an urban agricultural and leadership development organization. In this role, he helped the organization's leadership team create a "Six Domain Leadership Model." It offers human leadership lessons from studying the honey bee colony as a super successful organization. David earned a bachelor's degree in electrical engineering and a master's degree in business administration, with concentrations in finance and operations research. He worked in the computer industry for over 30 years.

www.ingramcontent.com/pod-product-compliance
Lightning Source LLC
Chambersburg PA
CBHW050257090426
42734CB00022B/3483